CONTENTS

5 **An Alien in the Backyard**

7 **An Unusual Bug**

13 **Eating and Escaping**

19 **The Next Generation**

22 Glossary

23 Find Out More

24 Index

24 About the Author

Praying mantises are common insects that you may have seen before.

An Alien in the Backyard

Walk outside and you'll see many insects. But wait! What is that? It's big. It's standing up. Uh-oh! It just turned its head to look at you. Is there an alien in your yard? No, it's a praying mantis.

The head of a praying mantis is triangular.

An Unusual Bug

Praying mantises look very strange. The largest can reach the size of a small bird. They can stand on their back legs. They hold their front legs as if they are praying. That is how they got their name. Their five eyes sit on a **triangular** head.

A mantis can turn its head from side to side.

The praying mantis can turn its head farther than any other insect. It can look from the far left to the far right. Then it might find a tasty snack. The insect only has one ear. It is found in the **thorax**. The praying mantis listens for bats who might want to eat it.

Most mantises are green or brown. They blend in with the leaves and grass.

This mantis is waiting for prey to come by.

Its colors allow it to sneak up on moths, crickets, grasshoppers, butterflies, and flies. Sometimes the insect captures bigger **prey**, such as lizards, frogs, or hummingbirds. Chomp!

Make a Guess!

Mantises eat a lot of different types of insects. Many farmers and home gardeners like it when mantises move into their fields and gardens. Why do you think they are happy to have mantises?

Praying mantises eat other insects, including butterflies.

Eating and Escaping

The long front legs of the mantis may make it look like it is praying. But it is really waiting for a snack to wander by. Then it grabs its prey with its front legs. The tiny spikes on those legs hold its prey in place.

These praying mantises are fighting each other.

An eating mantis should be left alone or there will be trouble. If another mantis approaches, the two may fight to the death.

Another bug stopping by to grab a nibble will regret it. The mantis shoots a thick, awful-smelling liquid out of its mouth. The liquid covers the bug's legs and **antennae**. The bug can't move and eventually dies.

This spider has caught a praying mantis to eat.

The mantis uses its legs for walking, climbing, jumping, and hunting. It also has two pairs of wings. It uses them to get away from **predators**. A mantis being chased by a bat will fly up and down, left and right to escape.

The word *mantis* means "**prophet**." Many **myths** surround the insect. Some say the mantis brings good luck. Others think it is some kind of alien. What is it about the insect that makes people believe these things?

This mantis is laying her ootheca on a leaf.

The Next Generation

When autumn begins, the female praying mantis lays her eggs. She might lay as many as 400 eggs on a twig or stem. She covers the eggs in a liquid that hardens. This keeps the eggs safe from hungry creatures and harsh winter weather. The egg case is called an **ootheca**. When spring arrives, the ootheca opens.

This mantis nymph climbing on a human finger shows how tiny they are.

Baby mantises look like adults, only smaller and without wings. They are called **nymphs**. The nymphs grow over the summer. They shed their outer skin, or **exoskeleton**, up to 10 times. In the fall, a new cycle begins.

There are more than 2,000 species of praying mantises. Twenty of those species live in the United States. The rest are in South America, South Africa, Europe, Australia, and southern Asia. Do you think the mantises in other countries look different from those in the United States?

GLOSSARY

antennae (an-TEN-ee) thin sensory organs on the heads of insects

exoskeleton (ek-so-SKEH-luh-tun) hard, protective covering

myths (MITHS) old stories that express the beliefs of a group of people

nymphs (NIMFS) a name used for some insects, including praying mantises, that have not yet become adults

ootheca (oh-eh-THEE-kuh) a praying mantis egg case

predators (PRED-uh-turz) animals that hunt, kill, and eat other animals

prey (PRAY) an animal that is hunted by other animals for food

prophet (PRAH-fit) a person who predicts the future

thorax (THOR-aks) the part of an insect's body between its head and its abdomen

triangular (trye-ANG-gyuh-lur) shaped like a triangle

FIND OUT MORE

BOOKS

Maley, Adrienne Houk. *20 Fun Facts about Praying Mantises.* New York: Gareth Stevens Publishing, 2013.

Penner, Lucille R. *Monster Bugs.* New York: Random House for Young Readers, 1996.

Roza, Greg. *Mysterious Mantises.* New York: Gareth Stevens Publishing, 2011.

WEB SITES

Audubon Magazine: Praying Mantis vs. Hummingbird
www.audubonmagazine.org /articles/birds/praying-mantis -vs-hummingbird
Watch this video of a praying martis trying to catch a hummingbird.

DesertUSA: Praying Mantis
www.desertusa.com/insects /praying-mantis.html
Read about the praying mantis species that live in the southwestern United States.

National Geographic: Praying Mantis
http://animals.nationalgeographic .com/animals/bugs/praying-mantis/
Read more about praying mantises and look at an amazing photo slideshow.

INDEX

A
antennae, 15

B
babies, 21

C
color, 9, 11

E
ear, 9
eating, 11, 12, 13–17
eggs, 19
escaping, 17
exoskeleton, 21
eyes, 7

F
fighting, 14, 15

H
head, 6, 7, 8, 9

I
insects, 4, 5, 11, 12

L
legs, 7, 13, 15, 17
life cycle, 19–21

N
nymphs, 20, 21

O
ootheca, 18, 19

P
predators, 9, 16, 17
prey, 10, 11, 13

S
size, 7

T
thorax, 9

W
wings, 17, 21

ABOUT THE AUTHOR

Tamra B. Orr is the author of more than 400 books for readers of all ages. Tamra, who has four grown kids, lives in the Pacific Northwest with her husband, dog, and cat—and perhaps, in the future, a praying mantis. She is a graduate of Ball State University and spends most of her free time reading, camping, and finding out more about the world around her.